Sunlight Sensations

Scott Wilken

Big Buddy Books

An Imprint of Abdo Publishing
abdobooks.com

abdobooks.com

Published by Abdo Publishing, a division of ABDO, PO Box 398166, Minneapolis, Minnesota 55439.

Printed in the United States of America, North Mankato, Minnesota
052025
092025

Design: Elena Klinkner, Mighty Media, Inc.
Production: Mighty Media, Inc.
Editor: Ruthie Van Oosbree
Cover Photograph: Lost_in_the_Midwest/Shutterstock
Interior Photographs: Andrei Stepanov/Shutterstock, p. 23; andrmoel/Shutterstock, pp. 20–21; artjazz/Shutterstock, p. 11; Atlantist Studio/Shutterstock, pp. 6–7; Cat Downie/Shutterstock, p. 17; Gergitek Gergi tavan/Shutterstock, pp. 8–9; Jamses/Shutterstock, pp. 24–25; jfkfoto.se/Shutterstock, pp. 12–13; Kanoktuch/Adobe Stock, p. 28 (cups); Kichigin/Shutterstock, pp. 26–27; Lasse Johansson/Shutterstock, p. 14; mayachitra/Shutterstock, p. 18; Michael Burrell/Adobe Stock, p. 28 (markers); Mighty Media Inc., p. 29; PhotoChur/Shutterstock, p. 5; thodonal/Adobe Stock, pp. 28 (paper towel)
Design Elements: Mighty Media, Inc.

Library of Congress Control Number: 2024948608

Publisher's Cataloging-in-Publication Data
Names: Wilken, Scott, author.
Title: Sunlight sensations / by Scott Wilken
Description: Minneapolis, Minnesota : Abdo Publishing, 2026 | Series: Weather wonders | Includes online resources and index.
Identifiers: ISBN 9781098296414 (lib. bdg.) | ISBN 9798384917847 (ebook)
Subjects: LCSH: Sun--Juvenile literature. | Sunshine--Juvenile literature. | Weather--Juvenile literature. | Sky--Juvenile literature.
Classification: DDC 551.6--dc23

Contents

Sunlight & Atmosphere

Earth gets light from the sun. This light passes through Earth's **atmosphere**. Sometimes water droplets and ice crystals in the atmosphere create amazing effects with the sun's light. The weather conditions have to be just right for these effects to occur.

Sun pillars are one amazing effect of sunlight hitting ice crystals in the atmosphere.

Prisms & Rainbows

Sunlight looks white. But it includes different colors. These are red, orange, yellow, green, blue, **indigo**, and violet. Sunlight bends when it shines through a **prism**. Each color bends a different amount. This separates the colors.

When the colors of sunlight separate, we can see each color individually.

When sunlight hits a **prism**, it bends and separates into different colors. The sunlight shines out as a rainbow. This process is called **refraction**. Ice crystals and water droplets act as prisms for sunlight.

Refraction causes rainbows.
Raindrops act as prisms for sunlight.

Reflected Light

In addition to being **refracted**, light can be **reflected**. This is when light hits a surface and bounces back rather than bending. Reflected light doesn't separate into different colors.

Mirrors work by reflecting light.

Sun Dogs

Sun dogs are bright spots in the sky next to the sun. Sun dogs most often occur during the winter. While rainbows need raindrops to form, sun dogs need ice crystals. So it must be cold enough for water in the **atmosphere** to freeze.

Ancient Greeks believed sun dogs were the god Zeus walking his dogs across the sky.

Sun dogs are also known as mock suns or parhelia.

The ice crystals in the **atmosphere** act like **prisms**. They **refract** the sun's light, forming spots next to the sun. They are a range of colors. The edges facing the sun are usually red, while the outside edges are usually blue.

Halos

Halos are rings of light around the sun or moon. Halos are also caused by ice crystals in the **atmosphere**. For a halo to form, there must be a thin cloud over the sun or moon. It must be cold enough for water in the cloud to freeze.

The clouds that form halos are usually 20,000 to 40,000 feet (6,096 to 12,192 m) high.

Moon halos (*pictured*) are fainter than sun halos. But they may be easier to see against the dark sky.

The ice crystals in the cloud **refract** light from the sun or refract light **reflected** off the moon. They also reflect the sunlight or moonlight. The **hexagonal column** shape of the crystals causes the light to form a ring around the sun or moon.

Tangent Arcs

Tangent **arcs** can occur with halos. They are also caused by sunlight **refracting** through ice crystals. In this case, the light forms an arc that bends away from the halo.

As the sun or moon rises, the arc widens.

Light Pillars

Light pillars are **columns** of light that seem to shoot up from bright objects. They appear when there are clouds with ice crystals in the **atmosphere**. The crystals are flat and **horizontal**. Bright light **reflects** off the crystals, creating light pillars.

Some reported unidentified flying object (UFO) sightings have turned out to be light pillars.

Light pillars can be caused by **artificial** light or natural light. When they are caused by the sun, they are called sun pillars. The moon and planets can also create light pillars.

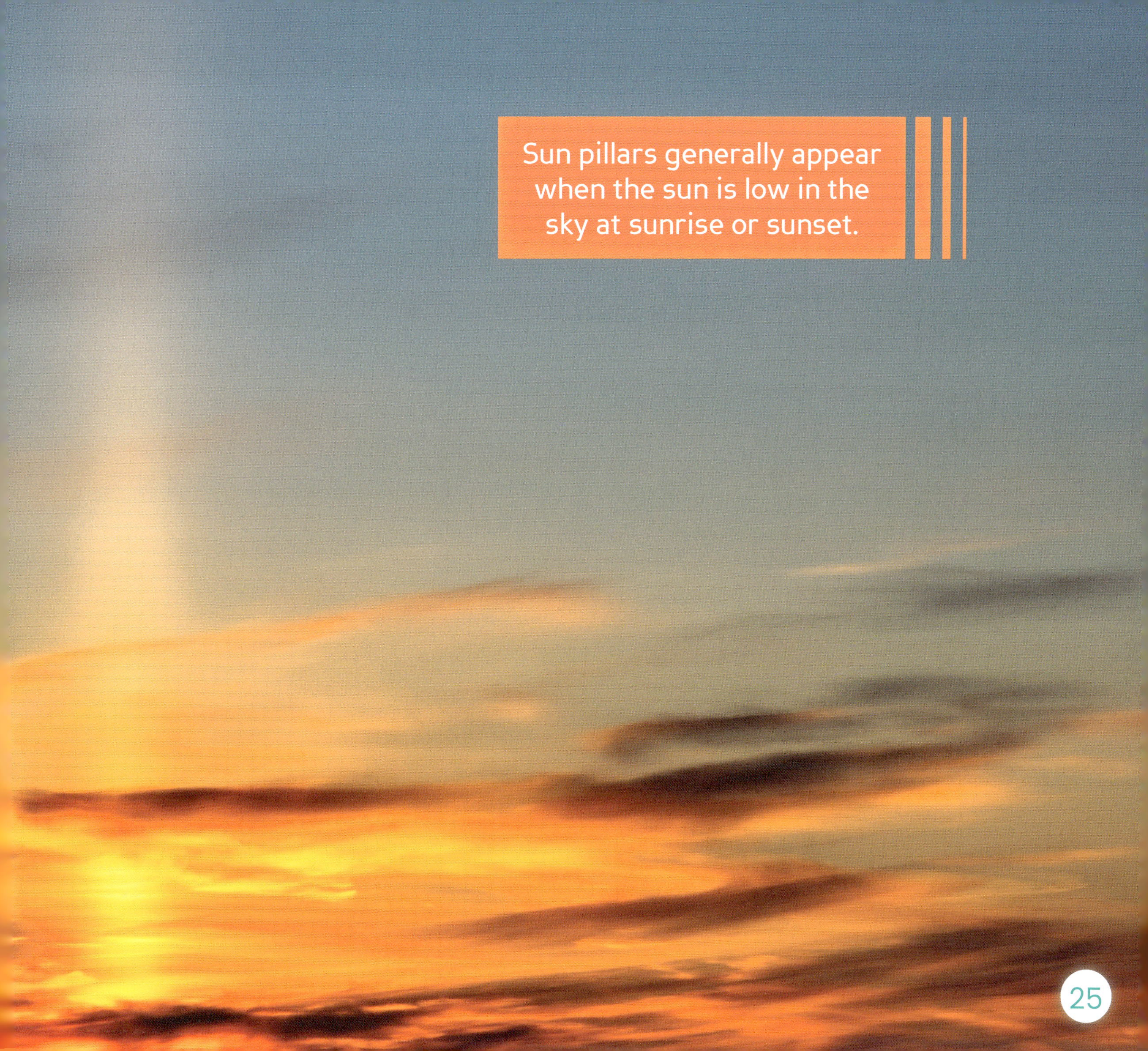

Sun pillars generally appear when the sun is low in the sky at sunrise or sunset.

Wondrous Weather

When elements in Earth's **atmosphere** combine with light, they can form amazing visual effects. If it's very cold out, look at the sky. Maybe you'll be lucky enough to see one of these weather wonders!

Light pillars take on the color of the light source causing them.

Grow a Rainbow!

What You Need

- paper towel
- two identical cups
- washable markers
- water

What You Do

1. Fold a paper towel so its ends will fit into the cups.
2. Color rainbow colors on the short edges of the paper towel. Make sure the colors on opposite ends of the paper towel match up. They should go in order: red, orange, yellow, green, blue, **indigo**, and violet.
3. Set each colored end of the paper towel in a cup of water. Wait for the paper towel to soak up the water to connect the rainbow!

Glossary

arc—a curved line.

artificial (art-uh-FISH-uhl)—made by humans.

atmosphere (AT-muhs-feer)—the layer of gases that surrounds a space object.

column—a straight, tall shape.

hexagon—a shape with six sides.

horizontal—level with the horizon, or side to side.

indigo—a purplish-blue color.

prism (PRIH-zuhm)—a clear object with angles that breaks light into a rainbow.

reflect—to throw back waves of light.

refract—to change the path of light rays as they pass from one space to another. Refraction is the process of changing the path of light rays.

Online Resources

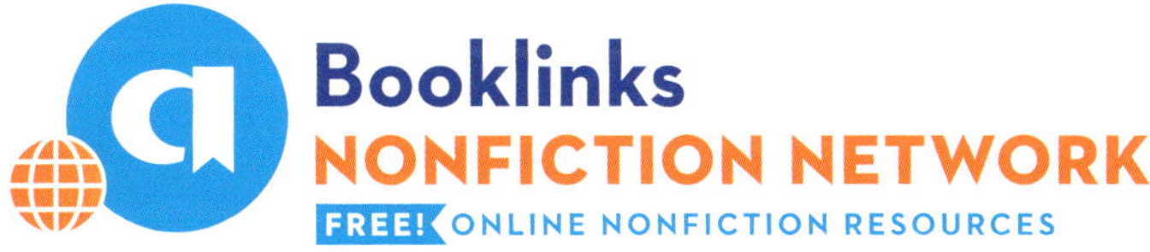

To learn more about sunlight sensations, please visit **abdobooklinks.com** or scan this QR code. These links are routinely monitored and updated to provide the most current information available.

Index